Rain Forests

Linda Aspen-Baxter

WEIGL PUBLISHERS INC.

Published by Weigl Publishers Inc.
350 5th Avenue, Suite 3304, PMB 6G
New York, NY 10118-0069
USA

Web site: www.weigl.com

Library of Congress Cataloging-in-Publication Data

Aspen-Baxter, Linda.
Rain Forests / Linda Aspen-Baxter.
p. cm. — (Biomes)
Includes index.
ISBN 1-59036-446-5 (hard cover : alk. paper) __
ISBN 1-59036-447-3 (soft cover : alk. paper)
1. Rain forest ecology—Juvenile literature. I. Title. II. Biomes (Weigl Publishers)
QH541.5.R27A87 2006 577.34—dc22 2006001032

Printed in China
1 2 3 4 5 6 7 8 9 0 10 09 08 07 06

Project Coordinator
Heather Kissock

Designers Warren Clark, Janine Vangool

Cover description: Rain forests cover about 10 percent of the Australian state of Tasmania.

All of the Internet URLs given in the book were valid at the time of publication. However, due to the dynamic nature of the Internet, some addresses may have changed, or sites may have ceased to exist since publication. While the author and publisher regret any inconvenience this may cause readers, no responsibility for any such changes can be accepted by either the author or the publisher.

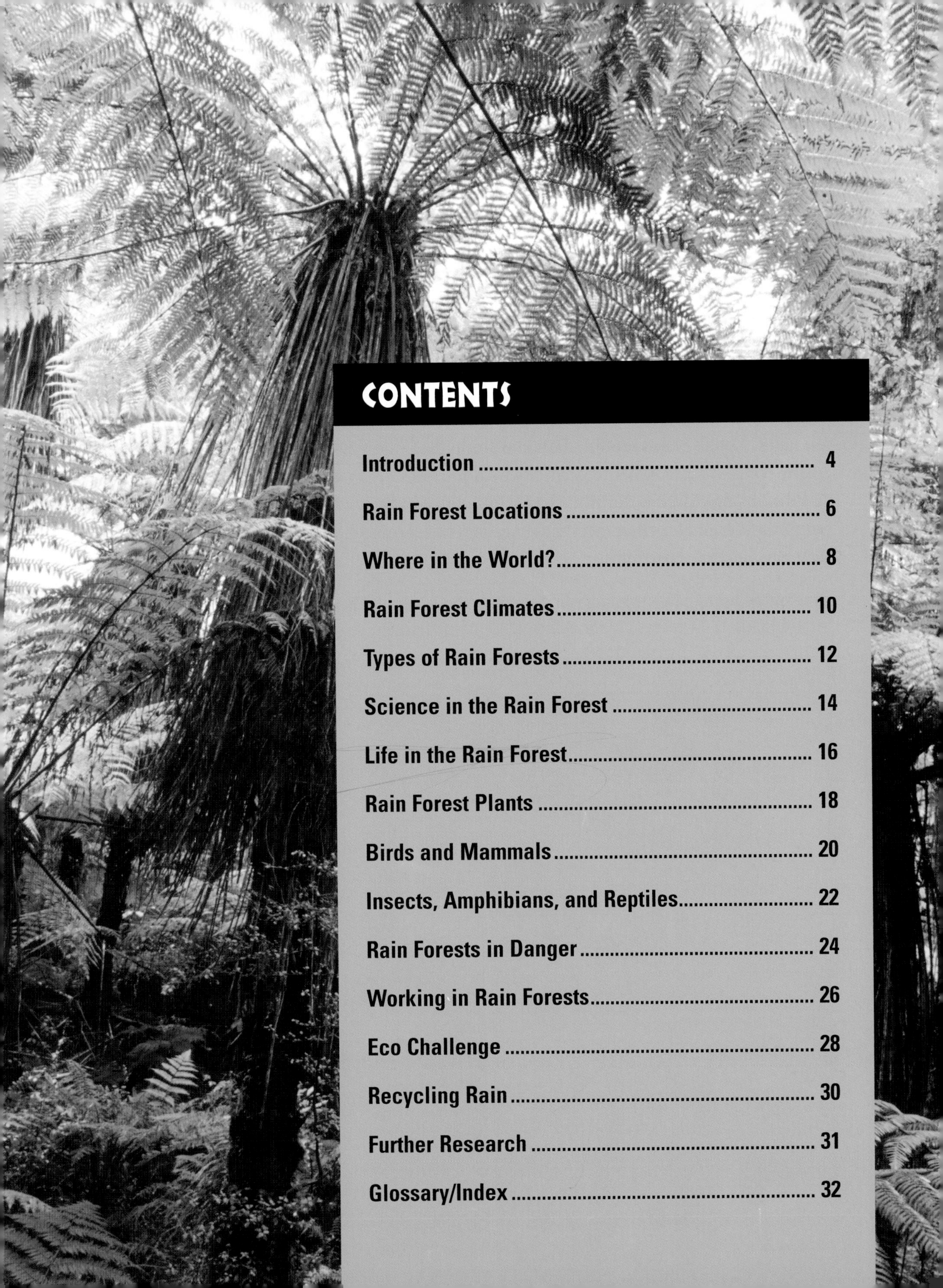

CONTENTS

Introduction

Earth is home to millions of different **organisms**, all of which have specific survival needs. These organisms rely on their environment, or the place where they live, for their survival. All plants and animals have relationships with their environment. They interact with the environment itself, as well as the other plants and animals within the environment. This interaction creates an **ecosystem**.

Different organisms have different needs. Not every animal can survive in extreme climates. Not all plants require the same amount of water. Earth is composed of many types of environments, each of which provides organisms with the living conditions they need to survive. Organisms with similar environmental needs form communities in areas that meet these needs. These areas are called biomes. A biome can have several ecosystems.

One of the best known tropical rain forests is the Amazon. This rain forest covers approximately 40 percent of the South American continent.

The Hoh rain forest is a temperate rain forest located in Washington's Olympic National Park.

Rain forests are known for their dense plant growth. The floor of a rain forest is often covered with thick vines and undergrowth. Tall trees rise straight up from the forest floor, and the treetops branch out to form a thick, green umbrella that covers the area below. This top layer is so thick that it blocks the sunlight.

There are an estimated 20 million plant and animal species in the rain forest. That is more than half of the plant and animal species on Earth. Most of these plants and animals grow and live in the trees. Many of them have adapted to life high off the ground.

Rain forests are forested biomes that receive high annual rainfall due to their location. There are two types of rain forest. Tropical rain forests are found near the equator where temperatures stay warm all year round. These rain forests receive rain almost every day. **Temperate** rain forests are found along sea coasts in the warm regions of the world. They receive their moisture from a combination of rain and fog.

FASCINATING FACTS

Rain forests account for about 40 percent of forests in tropical areas.

Tropical rain forests are Earth's oldest living ecosystems. Fossils from the forests of Southeast Asia indicate that these rain forests have existed for 70 to 100 million years. It is estimated that temperate rain forests can be up to 10,000 years old.

Rain Forest Locations

Rain forests are found in places that are very **humid**. Tropical rain forests are located in a broken band around the equator called the **tropics**. They are found in four main regions. The three largest of these regions are Central and West Africa, including Madagascar; Central and South America; and South and Southeast Asia. The fourth region is on the northern tip of Australia and Papua New Guinea. The largest rain forests are located in the following countries: Brazil in South America, Zaire in Africa, and Indonesia in Southeast Asia. Tropical rain forests are located in more than 40 countries around the world.

Temperate rain forests are found near large bodies of water, such as oceans and seas. The world's largest temperate rain forests are found along the Pacific coast of North America, stretching all the way from Alaska to Oregon. Other temperate rain forests are found along coastal areas of Chile, Great Britain, Norway, Japan, New Zealand, and South Australia.

The Daintree rain forest, in northern Australia, is one of the oldest rain forests in the world. It is estimated to be more than 130 million years old.

Rain forests in Costa Rica reach elevations of 9,000 feet (2,743 m) above sea level.

A rain forest can grow at almost any **elevation.** Some rain forests are found at more than 10,500 feet (3,200 meters) above sea level. Most rain forests, however, grow at elevations lower than 3,000 feet (914 m).

FASCINATING FACTS

The Amazon rain forest in South America is the largest tropical rain forest in the world. It covers more than 2.3 million square miles (6 million square kilometers). That is about two-thirds the size of the United States.

About fifty-seven percent of the world's tropical rain forests are found in Latin America.

WHERE IN THE WORLD?

Rain forest biomes are found on most of the world's continents. This map shows where the world's major tropical and temperate rain forests are located. Find the place where you live on the map. Which rain forest is closest to where you live? What type is it?

Tropical Rain Forest

Temperate Rain Forest

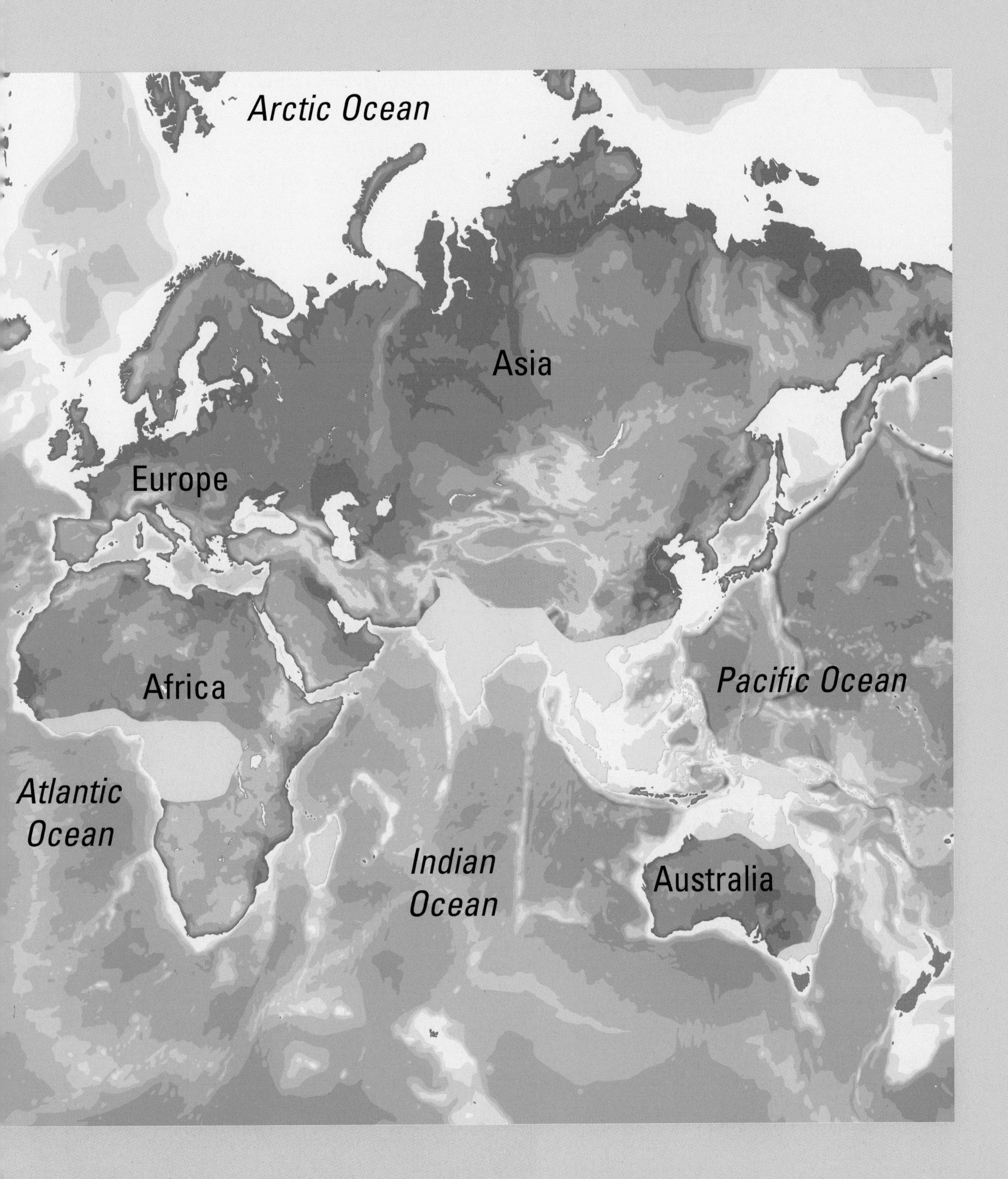
Arctic Ocean
Asia
Europe
Africa
Pacific Ocean
Atlantic Ocean
Indian Ocean
Australia

Rain Forest Climates

Rain forests have a warm, wet climate. There are mild temperatures year-round, and moisture falls almost every day. Even when it is not raining, the rain forest is very humid and damp. Average temperatures in a tropical rain forest range between 68° and 84° Fahrenheit (20° to 29° Celsius). A temperate rain forest will be cooler, but still warm. At least 80 inches (203 centimeters) of rain falls in a tropical rain forest every year, but many tropical rain forest areas receive up to 200 inches (508 cm) of rain. A temperate rain forest, on the other hand, receives between 60 and 200 inches of rain per year.

In a rain forest, the days are often cloudy with very light winds. Beneath the top layer of the trees, the air is still and moist. The moist, humid air insulates the rain forest, so it warms up and cools off slowly. This humid blanket of air keeps much of the Sun's energy from reaching the ground. At night, the humid air holds the heat close to the ground and keeps the forest warm.

Many plants found in the rain forest are unique to the biome. They are able to grow there because of the humid conditions.

FASCINATING FACTS

One year, the rain forest in Cherrapunji, India, received 1,042 inches (2,647 cm) of rain.

In the rain forests of South America, up to 250 billion tons (227 billion tonnes) of water vapor can be suspended in the air at any one time.

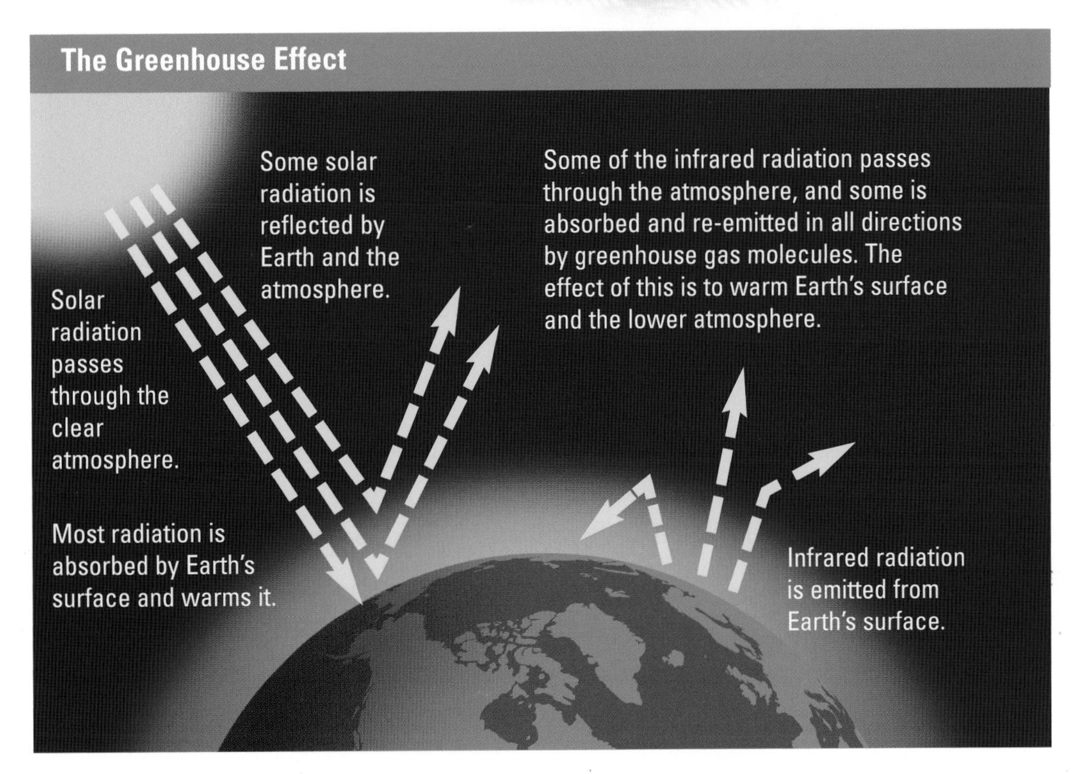

Rain forest climates impact the world as a whole. Due to the way they process moisture, rain forests affect weather patterns in other parts of the world. They also help keep the world's air clean and slow down the **greenhouse effect.**

Making Rain

Rain forests are often rain makers. This rain is made from the moisture of the rain forest. The moisture found within the forest evaporates into the air and forms clouds. These clouds move to other areas, where they eventually release rain. This rain feeds not only the rain forest, but other parts of the world as well. Rain cools the air, keeping temperatures from rising too high.

Air Conditioning

Rain helps the lush vegetation in a rain forest grow. These plants clean the air by absorbing carbon dioxide and turning it into oxygen through a process called **photosynthesis.** Carbon dioxide is a greenhouse gas. It helps keep Earth warm as part of the greenhouse effect process. However, when too much carbon dioxide is in the air, **global warming** can occur. Rain forests help regulate the amount of carbon dioxide in the atmosphere and assist in limiting the effects of global warming.

Types of Rain Forests

There are three main types of rain forests—lowland rain forests, montane rain forests, and cloud rain forests. Each forms at a different elevation.

Lowland rain forests are found at low elevations, such as the Amazon River basin in Brazil. In lowland rain forests, trees grow tall. There is little plant growth on the forest floor because little sunlight penetrates the thick cover of the treetops. At the forest edges, there is thick undergrowth. Most of the forest's moisture comes from rainfall.

Montane rain forests grow at altitudes higher than 3,000 feet (914 m), usually on the sides of mountains. These rain forests have a thick growth of trees that are usually shorter and have smaller leaves than those found in lowland rain forests.

Cloud rain forests are found on mountains at altitudes higher than 10,500 feet (3,200 m). They get their name from the low-lying clouds and mist that shroud them and keep them wet. The trees in a cloud rain forest are short and twisted. These forests receive enough light to promote the growth of creepers and mosses. Thick growths of mosses and ferns live on the branches and trunks of the trees, as well as along the forest floor.

Due to their higher altitude, cloud rain forests are much cooler than other rain forests.

FASCINATING FACTS

Mahogany trees are commonly found in lowland rain forests. These trees are known for their rich color and hard texture. They are used to make furniture and cabinets.

Pines, myrtles, laurels, and rhododendrons are just some of the plants found in montane rain forests.

Rain forests form layers of life. Different types of plants and animals live at different layers of the rain forest.

Layers of a Rain Forest

Emergent Layer The tops of some of the tallest rain forest trees stick up above the main "roof," or canopy layer, of the forest. These trees can reach heights of 100 to 250 feet (30 to 76 m).They form the emergent layer.

Canopy Layer The canopy forms the roof of the rain forest. Trees ranging in height from 98 to 164 feet (30 to 50 m) form this thick layer. In the rain forest, the emergent and canopy layers receive the most light.

Understory Layer Under the canopy are trees with thinner trunks and narrow crowns. They form the understory layer of the forest. This layer receives very little light. Plants in this layer rarely grow very large.

Forest Floor Less than one percent of the light received by the canopy reaches the forest floor. There is little plant life here. Fungi, ants, earthworms, termites, and bacteria live among the tree roots. They feed on rotten leaves, fruit, animal droppings, and branches that fall from above.

Science in the Rain Forest

Rain forests provide the world with clean air and help moderate temperatures. They also play a key role in determining weather patterns. Scientists are constantly studying what happens inside rain forests and how this biome affects the world and the people in it. Studies have not only focused on how to preserve rain forests, but on how to properly develop the areas in which these biomes are found.

Trees are the focal point of a rain forest. There are many uses for trees, and, in the search for more wood, forestry companies have entered rain forests and removed large numbers of trees. Trees are important in the absorption of carbon dioxide. When they are cut down, the world loses some of its ability to maintain temperature patterns. Scientists are researching the degree to which logging is affecting weather and what can be done to correct the problem.

Carbon dioxide can be either absorbed or created, depending on the type of vegetation and its location in the forest. To determine which parts of a rain forest absorb carbon dioxide and which create it, scientists have used an airplane equipped with chemical sensors to measure levels of carbon dioxide in different parts of the Amazon rain forest. This kind of study may help decide where human development of the rain forest can take place.

The removal of trees during the building of Panamerican highway in South America impacted global warming patterns.

Satellite imagery shows that, besides logging, the Amazon rain forest is also being burned to make room for farmland.

Scientists are also using satellites to view the areas that are being logged. They are determining how much of a rain forest is left and how much damage logging has caused to the forest's canopy. This information helps scientists suggest ways to promote forest regrowth.

When a rain forest's canopy is damaged, the vegetation below is also affected. This concerns scientists because, in recent years, they have found plants in the rain forest with medicinal properties. It is estimated that rain forests are providing sources for one-quarter of the world's medicines. Many of these medicines are being used in cancer treatments. Plants from rain forests all over the world are being taken to laboratories for analysis. Preserving the rain forest biome will not only assure the supply of current medicines, but possibly a supply of medicines yet to be discovered.

FASCINATING FACTS

About a quarter of the medicines used today come from rain forest plants. Curare is used as an anesthetic and to relax muscles during surgery. It comes from the bark of a tropical vine. The rosy periwinkle is used to treat leukemia.

Scientists estimate that there are at least 30,000 plant species that have yet to be discovered. Most of these species are believed to be in the world's rain forests.

LIFE IN THE RAIN FOREST

More than half of Earth's estimated 10 million species of plants, animals, and insects live in tropical rain forests. To date, scientists have studied less than one percent of the millions of species in the rain forest.

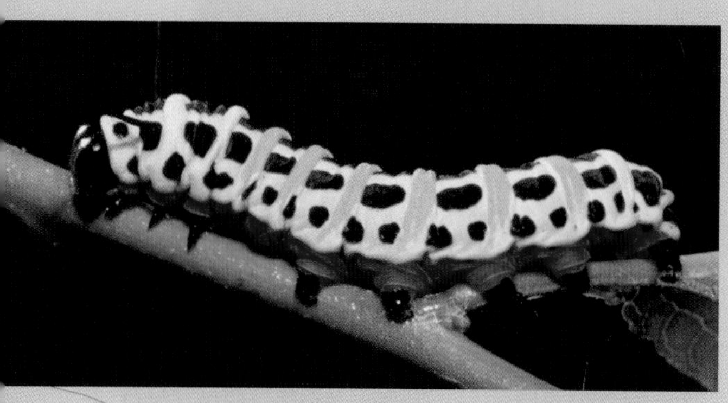

A caterpillar grows to be about 27,000 times larger than it was as an egg.

INVERTEBRATES

The rain forest is ideal for **invertebrates** with soft bodies, such as leeches and caterpillars. The humid air keeps them from drying out. Many invertebrates live on the forest floor, in streams, or in pools of rainwater. They eat plants or decaying animal matter. Insects are just one type of invertebrate, but they make up almost 90 percent of all animal life in tropical forests. They include butterflies, mosquitoes, ticks, and huge colonies of ants. Bees, butterflies, and moths gather pollen and nectar from the many flowers and other small plants in this biome.

REPTILES AND AMPHIBIANS

The tropical rain forest is a perfect environment for reptiles, such as snakes and lizards, and amphibians, such as frogs and salamanders. Both are cold-blooded, so their bodies are about the same temperature as their environment. In the rain forest, their body temperature stays warm, so they can stay active all year. Many reptiles and amphibians have patterned or colored skin that **camouflages** them against the forest background. Long tongues with sticky tips help some reptiles and amphibians catch insects. Others eat small mammals and birds.

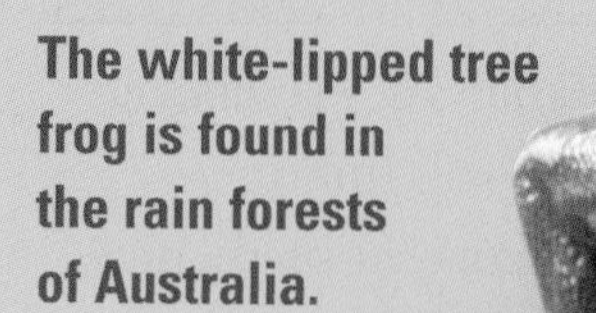

The white-lipped tree frog is found in the rain forests of Australia.

The rufous-tailed hummingbird is found in rain forests ranging from Mexico to Colombia.

BIRDS

More birds live in tropical rain forests than in any other biome. Most do not need camouflage to protect them, so their feathers are brightly colored. They have developed short, broad wings because they do not have much room for flying in the dense forest. Different birds feed in each of the layers of the rain forest. Parrots hunt for seeds and insects in the canopy. Pittas dig around on the ground for snails and ants. Hummingbirds feed on nectar and insects in flowers.

MAMMALS

Most mammals in the rain forest have adapted to life high in the trees. In the canopy, small mammals often eat plants and insects. The agouti, paca, and royal antelope are herbivores that feed on leaf buds and fruit. Sloths and colobus monkeys are also leaf-eaters. Members of the cat family, such as the palm civet and ocelot, are carnivores. Orangutans are omnivores. They eat plants and animals. Monkeys, kinkajous, coatis, bats, okapis, gorillas, and jaguars are also found in the rain forest. Primates, such as pygmy marmosets, bush babies, and lemurs cling to tree trunks. They gnaw away bark so they can drink tree sap.

Mountain gorillas are found only in the rain forests of Rwanda, Uganda, and Zaire.

Rain Forest Plants

Trees

Trees in rain forest biomes grow almost continuously because of the warm temperatures and constant rainfall. Rain forest trees rarely grow more than 200 feet (61 m) tall. Most have a single, straight trunk that grows upward to reach the light at the canopy. Cinchona and cacao are common trees in a tropical rain forest. The cinchona tree is important for its bark. Several medicines are made from it, including quinine, which is a treatment for malaria. The seeds of the cacao tree of Central and South America are used to produce chocolate. The Sitka spruce and western hemlock are often found in temperate rain forests. Both are good lumber trees.

Lianas can be small vines that grow up along a tree trunk or large vines that are as thick as the trunk itself.

Climbing Plants

Climbing plants use hook-like tendrils to climb up the trunks and along the limbs of trees to reach light at the canopy. Lianas are climbers found in all tropical rain forests. They look and grow like vines, but they have sturdier, woody tissues like trees. When lianas reach the canopy, they drape themselves on the branches. Lianas grow leaves and branches. They develop flowers and fruit. Lianas even send out roots that dangle in the air and collect nutrients.

Sitka spruce and western hemlock are common trees in the temperate rain forests of North America's northwest coast.

Mosses can be found growing on rocks, soil, and the bark of trees.

Epiphytes

Epiphytes are also known as "air plants." These are plants that grow on other plants. They do not send any roots to the ground. Instead, their roots are exposed to the air. Their roots absorb the nutrients they need from air and forest litter. Epiphytes grow on trees and other plants in the canopy. Ferns, lichens, and orchids are epiphytes that live on tree branches in the rain forest.

Moss

Moss is a small, feathery plant that grows on the forest floor and in bogs. In cool, moist places with acidic soil, moss can grow several feet (1 to 2 m) thick. Moss can absorb huge amounts of moisture. Sphagnum moss can hold up to 20 times its weight in water.

FASCINATING FACTS

In just one 25-acre (10-hectare) patch of Malaysian rain forest, scientists found 750 different tree species. That is more than all the native tree species in the United States and Canada.

Rain forest trees have shallow root systems. Many trees develop plates of very hard wood called buttresses to give them a more stable base. The buttresses grow out from the bottoms of trunks above ground. They reach as much as 30 feet (9 m) up the trunk and spread out about the same distance along the ground.

The leaves of many plants and trees in the rain forest biome have a slick coating and drip tips. These narrow points at the end of the leaves allow rain to run off easily.

Birds and Mammals

Birds

Most rain forest birds live in the canopy and have adapted to life high above the ground. The harpy eagle lives in the forest canopy of Central and South America. It has very sharp eyes, so it often sits in an emergent tree watching for prey. More than 100 species of parrots live in rain forests. These brightly colored birds have very strong feet that allow the birds to hang upside down from a branch for long periods of time. Junglefowl live mostly on the forest floor. They are the ancestor of the chicken. The males crow like roosters and even have a bright red comb on top of their heads.

Sloths are the slowest mammals on Earth.

A harpy eagle's talons can be 5 inches long (12.5 cm), the same length as a grizzly bear's claws.

Sloths

Sloths are found in the rain forests of South America. They spend their lives hanging upside down from tree limbs and almost never leave the trees. Sloths cannot stand or walk on the ground. They use their long claws to grip the trees. Their fur even grows the opposite way to fur on most mammals. Rainwater runs off easily as the sloth hangs upside down.

Primates

Rain forests are home to many species of primates. Gorillas live in family groups in the rain forests of Africa. On the ground, gorillas normally move in a stooped posture, with the knuckles of their hands resting on the ground. Females and young gorillas climb trees, but males rarely climb because of their size. Gorillas rise early and feed on leaves, buds, stalks, berries, bark, and ferns. Then they relax and rest. They feed again in the afternoon. Orangutans walk on all fours or erect on two legs. Sometimes they swing by their hands from branch to branch. They feed mostly on fruit, but will also feed on leaves, seeds, young birds, and eggs. The orangutan sleeps in the trees in a platform nest made of sticks. Chimpanzees also live in the rain forests of Africa. Chimpanzees have been seen sticking plant stems into termite nests to drive the termites out so they can eat them. They even use leafy tree branches as umbrellas during heavy rains.

FASCINATING FACTS

Sloths spend up to 18 hours a day hanging in the trees without moving.

The world's smallest mammal is about 1 inch (2.5 cm) long. It is the Kitti's hog-nosed bat, and it lives in the rain forests of Thailand.

Some mammals live on the forest floor. The okapi is related to the giraffe, but it is smaller and has a shorter neck. It has zebra-like stripes on its hindquarters and upper legs. The okapi lives in the rain forests of the Congo in Africa. It has a long, sticky tongue, which it uses to forage for leaves, fruits, shoots, and berries.

Insects, Amphibians, and Reptiles

Walking sticks can regenerate legs that fall off.

Insects

Insects have developed ways to survive in the rain forest. They use camouflage to hide from insect eaters. Katydids, walking sticks, and moths look like twigs, leaves, or bark. Some tropical leafhoppers look like sharp thorns on a plant's stem. Butterflies may look like dead leaves when their wings are closed. When they are threatened by a predator, they open their wings and flash bright colors. This startle defense gives the butterfly time to escape. The ants and treehoppers of Central America help each other survive. The ants guard the treehoppers as they drink juices from plant stems. The ants eat the sugary droppings of the treehoppers.

Amphibians

Tropical rain forests are home to tree frogs and poison arrow frogs. Tree frogs have large eyes that help them see well in the dim light of the understory. Some species have suction pads on their toes that secrete sticky mucus. This helps them cling to tree trunks and branches. The poison arrow frog can be yellow, red, or blue. The color warns predators that the frog is poisonous. Poison arrow frogs lay their eggs on wet leaves and guard them until they hatch. Then one of the parents carries the tadpoles on its back to tiny pools in leaves up in the trees. It then puts one tadpole in each mini-pool. After about two months, the tadpoles become frogs, and they move out of the water.

Most poison arrow frogs live in the leaf litter found on the rain forest's floor.

Reptiles

Tropical rain forests are home to a variety of reptiles. Three of the most well-known rain forest snakes are West Africa's vine snake as well as the bushmaster and fer-de-lance of Central and South America. The vine snake is about as thick as a man's finger. Its long, slender body and greenish-brown color help the vine snake blend in with the creepers, vines, and branches of the forest. The bushmaster and fer-de-lance are pit vipers. Pits or indentations behind and above their nostrils help them detect the slight temperature changes caused by warm-blooded mammals. Both snakes kill their prey with venom. The Parson's chameleon of Africa and Indonesia's Komodo dragon are two well-known rain forest lizards. The Parson's chameleon is a pro at camouflage. It lives in the canopy and can change its color to look just like the leaves around it. Sometimes it even trembles to mimic the leaves moving in the breeze. The world's largest lizard is the Komodo dragon of Indonesia. It can grow to 10 feet (3 m) in length, and it can weigh up to 300 pounds (136 kilograms).

FASCINATING FACTS

South American Indians hold poison arrow frogs over a flame until poison oozes out of the frogs' skin. Then the hunters dip their arrows in the poison.

The fer-de-lance is the most dangerous snake of Central and South America. One bite injects twice the fatal dose of venom for humans.

The Komodo dragon has deadly bacteria in its mouth. After it bites its prey, the prey becomes sick and dies from blood poisoning within a day or two. Then the Komodo dragon eats the animal.

Rain Forests in Danger

Tropical rain forests once covered 14 percent of Earth's land surface. Now they cover only six percent. Huge areas of rain forest have been destroyed by logging and clearing of land for farms and ranches. Since 1940, most of the rain forest in Southeast Asia has been destroyed as a result of these activities.

As populations grow, more land is needed for farms and cattle pasture. In developing countries, "slash and burn" agriculture is practiced. Trees are cut, and the undergrowth is burned to clear land for farms. As soon as the burned ground cools, crops are planted. The ash from the burning gives the crops nutrients that help them grow quickly, but these nutrients are used up quickly. Farming in these cleared patches only lasts two to seven years. Then the farmer abandons the land and moves to a new area where another patch of rain forest is cut down and burned.

It is estimated that 25 million acres (10 million ha) of tropical rain forest are cleared every year for agricultural purposes.

Commercial logging is the single largest cause of tropical rain forest destruction.

Logging wood from rain forests for furniture and building materials is a lucrative business. Clear-cutting, a logging technique in which all of the trees in an area are removed at once, is one of the most common ways for logging companies to fell trees, but it destroys habitat and disrupts normal forest growth. Often, cleared areas are not replanted, so the rain forest cannot restore itself. Rain forests help conserve water and protect the soil from erosion. When the forests are gone, heavy rains wash the soil away, and flooding happens more often.

The rapid rate of rain forest destruction is causing the loss of 1,000 or more orangutans yearly in Sumatra alone.

The animals of the rain forest are losing their habitats because of deforestation. Several species of rain forest animals are now extinct. Many more are endangered. Orangutans and gorillas are just two of these animals. Orangutans live in areas with deep forest cover, and their habitat is quickly disappearing. These primates used to live all over Southeast Asia and on the islands that stretch between Asia and Australia. Today, they are found only in a few remaining pockets of forest on the islands of Borneo and Sumatra. A similar situation exists with the mountain gorilla. There are currently only about 600 mountain gorillas left in the world, and they live in two populations that are separated by areas of human development. Due to deforestation, they are no longer able to wander freely within their home range.

WORKING IN RAIN FORESTS

People who work in rain forests play an important role in increasing awareness about this biome and the issues facing it. They find ways to improve and protect rain forest habitats. These people also focus attention on the vital role the rain forest biome plays in the global environment.

BIOLOGISTS

- Duties: studies the plant and animal life that live in the rain forest biome
- Education: bachelor of science in biological science or environmental biology
- Interests: animals and plants, chemistry, environment, working outdoors, math, science, conservation

Biologists study the plant and animal life that live in rain forests and anything that affects the natural balance of the rain forest biome. Biologists count populations of rain forest life at regular intervals to find out if the numbers are increasing or decreasing. They also measure plant **biomass** to assess the differences between environments. Biologists work on solutions to improve the health of the rain forest biome.

RESEARCH SCIENTISTS

- Duties: collects and records data in rain forests; studying the plants, animals, and climates of the rain forest
- Education: bachelor of science degree
- Interests: Earth science, ecosystems, science, geology, biology

Research scientists identify and classify new plant and animal species in the rain forest. They study life in the layers of the rain forest and investigate how to grow different species of rain forest plants for future reforestation projects. Scientists perform research in the tropical rain forest to find possible drugs and cures for various illnesses. They work with rain forest tribal **shamans** and herbal healers to learn what they know about the plants of the rain forest and their many uses.

ENVIRONMENT CONSULTANTS

- Duties: studies environments and determines ways to protect them
- Education: bachelor's degree in environmental design or natural resource management
- Interests: environment, nature, conservation

Environment consultants study the ways in which pollution and human activity affect the rain forest biome. They look for ways to preserve rain forest areas, and the plants and animals that live in them. They work to have rain forests declared protected parks and reserves.

ECO CHALLENGE

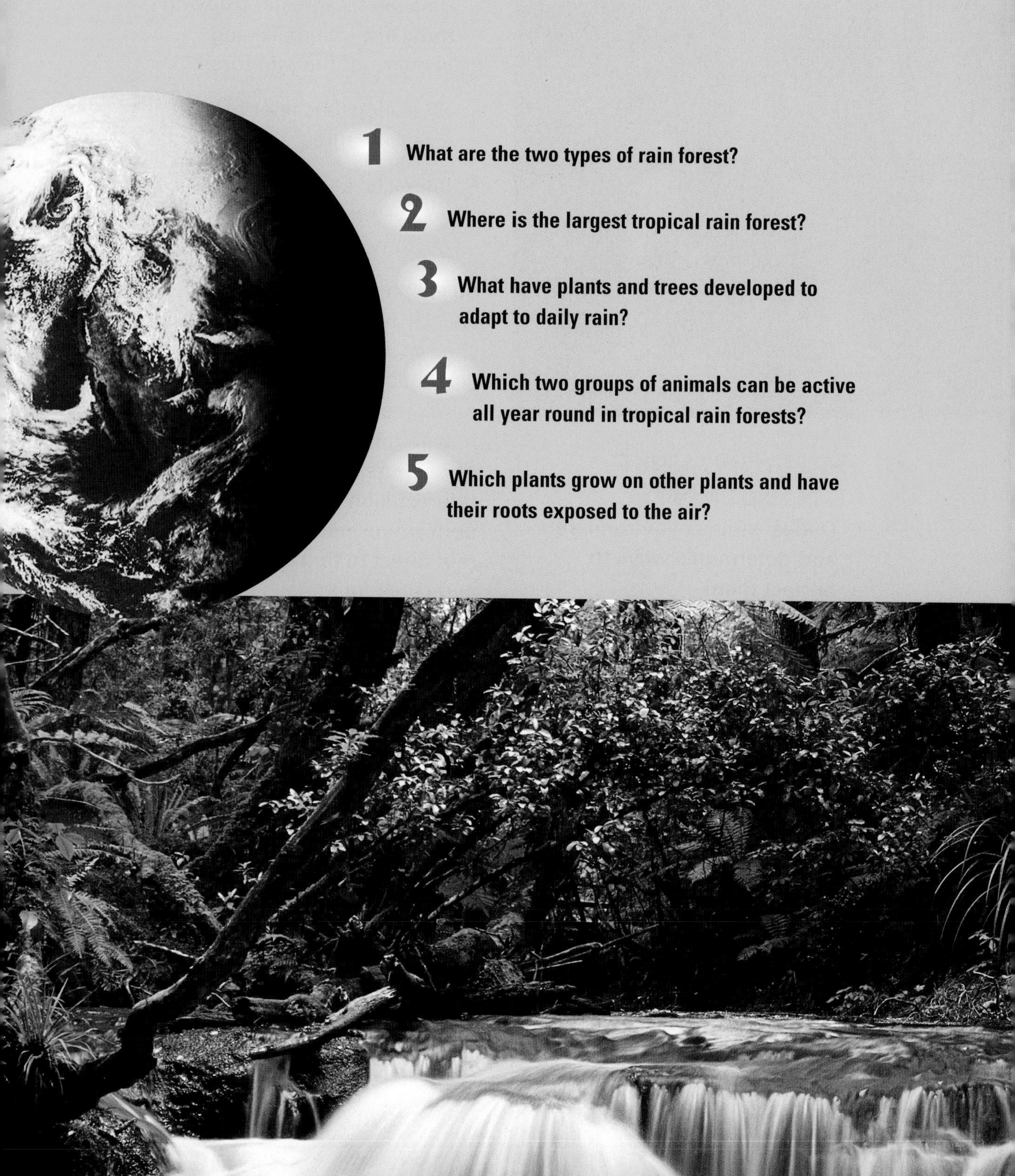

1. What are the two types of rain forest?
2. Where is the largest tropical rain forest?
3. What have plants and trees developed to adapt to daily rain?
4. Which two groups of animals can be active all year round in tropical rain forests?
5. Which plants grow on other plants and have their roots exposed to the air?

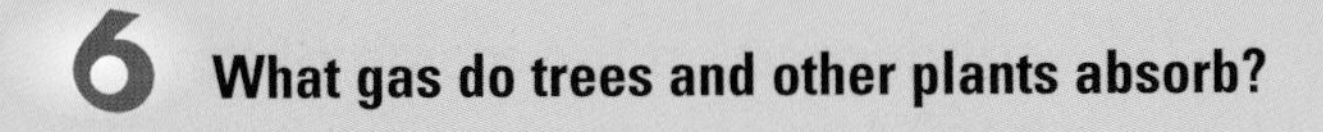

6 What gas do trees and other plants absorb?

7 How much of Earth's land surface do rain forests cover today?

8 What type of rain forest grows at the highest elevation?

9 How much light reaches the forest floor?

10 What helps tropical rain forests warm up and cool off slowly?

Answers

1. tropical and temperate
2. Amazon basin in South America
3. drip tips and a slick coating on leaves
4. reptiles and amphibians
5. epiphytes and lianas
6. carbon dioxide
7. six percent
8. cloud rain forest
9. less than one percent
10. moist, humid air

RECYCLING RAIN

With this activity, you can see how rain is recycled in tropical rain forests.

MATERIALS

- clear, plastic bag
- tree branch
- sunny day
- twist ties

1. Tie a clear, plastic bag around the leafy end of a tree branch in a sunny place. Use the twist tie to close the ends of the bag snugly around the branch.
2. Wait one hour. Then return and look at the bag.

Do you see droplets of water condensed on the inside of the bag? That water came from the leaves on the branch. Leaves lose water through their cell walls and through tiny holes called stomates. This is called transpiration. When this water evaporates, it forms clouds that rain back down on the rain forest.

Try the bag test on different kinds of plants. Try cacti, leafy trees, shrubs, and flowers. Which ones give off more water?

FURTHER RESEARCH

How can I find more information about ecosystems, rain forests, and animals?

- Libraries have many interesting books about ecosystems, rain forests, and animals.
- Science centers and aquariums are great places to learn about ecosystems, rain forests, and animals.
- The Internet offers some great websites dedicated to ecosystems, rain forests, and animals.

BOOKS

Albert, Toni. *The Remarkable Rainforest.* Mechanicsburg, PA: Trickle Creek Books, 2003.

Johnson, Rebecca L. *A Walk in the Rain Forest.* Minneapolis, MN: Carolrhoda Books, Inc., 2001.

Knight, Tim. *Journey into the Rainforest.* New York, NY: Oxford University Press, 2001.

WEBSITES

Where can I learn more about rain forests and other biomes?

What's It Like Where You Live? – Biomes of the World
http://mbgnet.mobot.org/

How can I learn more about rain forests?

How Stuff Works—How Rainforests Work
www.howstuffworks.com/rainforest.htm

Journey into Amazonia
www.pbs.org/journeyintoamazonia/about.html

Rainforest Live
www.rainforestlive.org.uk/

GLOSSARY

biomass: the weight of roots, shoots, leaves, tree trunks, and other plant material that exists in a certain area

camouflages: uses protective coloring to blend into natural surroundings

ecosystem: a community of living things sharing an environment

elevation: the height to which something rises

global warming: an increase in the average temperature of Earth's atmosphere; enough to cause climate change

greenhouse effect: the trapping of heat from the Sun within Earth's atmosphere

humid: damp, moist

invertebrates: animals that do not have a backbone

organisms: living things

photosynthesis: the process in which a green plant uses sunlight to change water and carbon dioxide into food for itself

shamans: tribal elders

temperate: a climate that is usually mild without extremely cold or extremely hot temperatures

tropics: a region of Earth that is near the equator

INDEX